The Great Compost Heap

Text copyright © 2014 by Renaee Smith
Illustrations copyright © 2014 by Somnath Chatterjee

Author: Renaee Smith
Illustrator: Somnath Chatterjee

Printed in the United States of America

Acknowledgements

I would like to thank my family and friends for believing in my vision.

I would like to thank Dr. Ordel Brown for encouraging me and for reviewing my manuscript.

ISBN-13: 978-0-9855415-1-4

*F*or Sheldon, Reuben and Abraham

For the Teacher or Parent

In The Great Compost Heap the student is introduced to the idea of composting and how composting helps to further the idea of the 3 R's, "Reduce, Reuse, and Recycle".

For the Teacher – starting a classroom compost heap/bin is a great way to teach students about recycling and the environment. This is also an activity that all the students in the class can participate in. The students will be able to put leftovers from their lunch like banana peels and apple cores in the compost bin. This activity can be started at the beginning of the school year and can be used as a means to discuss a variety of topics such as the food chain, how things decay and ultimately using the compost to fertilize a class or school garden.

For the Parent – starting a compost heap at home can be a very exciting adventure for children of all ages. You will have the benefit of watching your kitchen waste breakdown. It will be an opportunity for the whole family to get involved in seeing how much they can keep out of the landfill. In a few months your lawn and garden will benefit from the compost.

The book shows the student how to create their own compost heap whether at home or school using simple items. This book will guide the student and help him/her to observe the process and at the end perform an experiment, to test the quality of the compost.

At the end of the book you will find a word wall, this list words that were used in the book that might be difficult for students to read. There is also a crossword puzzle for the students to do just for fun.

The Great Compost Heap

By Renaee Smith

Illustrated by Somnath Chatterjee

"Hi mom," Freddie said as he rushed into the kitchen and started rummaging in the fridge. "Hi Freddie," said Freddie's mom, smiling. "How was your day?" she asked.

Freddie's mom always asked him about school. He told her about his classes and what he learned in each one. Then he told her about what he did in science class. Science was Freddie's favorite subject.

He told his mom that Mr. Fran, his science teacher, talked about "Reduce, Reuse and Recycle". He told us we should always try to recycle our garbage so we can have a cleaner environment.

He talked about what happens when we throw out our garbage. When we throw out our garbage it gets taken to a huge dump and everything gets mixed together.

The dump gives off a lot of harmful gases that are not good for the environment. Even though things like our fruit peels and vegetable peels are biodegradable they take a very long time to break down when they are mixed together with other waste. When everything gets packed together, there is not enough air around to help the fruit peels and vegetables decompose.

"Mom, the interesting thing that I learned today, was about making a compost heap. A compost heap will help us to recycle and reduce those harmful gases being released into the air.

Mr. Fran told us that a compost heap is a pile of garden and kitchen waste that decomposes to produce a rich soil which can add nutrients and improve the quality of the soil used in gardening. Composting is nature's way of recycling.

"Mom did you know what a compost heap is?" asked Freddie. "No, I did not," replied Freddie's mom.

"You know mom; I want to make my own compost heap so I can help the environment," said Freddie.

"What do you need to do to create this compost heap?" said Freddie's mom.

"My teacher said we only need a few things: a big bin or box, fruit peels, vegetable peels things that we throw out, but not meat or dairy that is things like milk and cheese. We can put leaves and twigs and some soil, which are all things we have already. This is going to be so cool! I can't wait," said Freddie.

The next day Freddie, his mom and his dad went in the back yard to make the space for the compost heap.

They built a big box and put a layer of soil on the bottom, then, they layered in the items they had collected from the kitchen like banana skins, apple core and potato skins. Outside in the yard they collected leaves, twigs and some grass and put that in the box. Then they added another layer of soil.

Freddie was so excited, he told everyone that passed his home about his compost heap and how he was doing his part to save the earth.

Everyone started to get excited as well, and by the time Freddie and his parents were done all the neighbors wanted to help with the compost heap.

Every day the neighbors would drop off their collections at Freddie's house so he could put it in his compost heap.

"Mom and dad, look how big the compost heap is getting," said Freddie. Every day Freddie would go in the backyard and mix the heap so the peels and leaves would mix well. He would add more soil as he got more peels.

Very soon the compost heap became really big as more people brought in their collections. Eventually Freddie and his mom had to make the box bigger.

Pretty soon Freddie and his mom did not have enough room in the box to put all the peels they were collecting, and they needed a bigger space.

The mayor of the town, Mayor Folley, heard about the interesting project that Freddie and his mom were working on. He decided to create a special place in the town for everyone to drop off all their vegetable peels and fruit peels.

THE COMPOST

The people in the town dropped off all their collections at the compost heap, and the pile got bigger and bigger. Mayor Folley put Freddie in charge of keeping the compost heap healthy.

The peels were collected for several months. Over that time Freddie made sure that the compost had the right amount of water and that the right things were put in there.

Freddie told Mayor Folley what Mr. Fran had said - that the soil from the compost heap could be used to fertilize the vegetable gardens and the plants.

Mayor Folley made a big announcement that anyone who had a garden could stop by the compost heap and collect some soil.

Freddie was so excited that the people of the town could use the compost to help to fertilize their crops, and they had bigger fruits and vegetables than they had ever had before.

The tomatoes and carrots were the biggest they had ever seen. The roses and tulips were the largest and most beautiful they have ever seen.

They were so happy that Freddie told them about composting, because not only were they helping to save the environment, they were also getting the best harvest that they ever had.

Freddie's compost heap soon became known as The Great Compost Heap.

THE GREAT COMPOST HEAP

How to Make a Compost Heap

How to Make a Compost Heap

OBJECTIVE: To learn about composting

MATERIALS:
Wooden box or Storage bin with cover
Soil
Shovel

Kitchen Compost

• Vegetable peels	• Egg shells
• Fruit peels	• Nut Shells
• Coffee grounds	• Vegetable or fruit scraps

Note: Do not add meat scraps, bones, dairy products, oil or fat

Garden Compost

• Grass Clippings	• Sawdust
• Leaves	• Wood Chips
• Ashes	• Weeds and other garden waste
• Shredded Paper	

PROCEDURE:

1. Get a wooden box or plastic bin.

2. Spread a layer of soil in the bottom of the bin.

3. Place kitchen or garden wastes into the composting bin. Make sure to cut up the materials into small pieces so they can decompose quickly.

4. Spread another layer of soil over the pile. This helps to keep the surface moist as well as bring in the worms that help in making the compost.

Note: If the compost heap gets too moist add dry leaves or shredded paper, or add water to the heap if it gets too dry. The compost heap should be damp to the touch, but not wet that drops come out when you squeeze it.

OBSERVE:

1. Allow the heap to "cook". It should heat up quickly and reach the desired temperature (90 to 140 degree F, or 32 to 60 degree F) in four to five days.

2. Stir your compost every 2 to 3 days by turning it with a shovel if you want to speed up the time it takes for the kitchen and garden waste to break down.

3. The pile will settle down from its original height. This is a good sign that the compost is progressing properly.

4. Stir your compost heap every week, it should be ready to use, in one to two months. If you don't turn it, the compost won't be ready before six to twelve months.

5. Your compost should look like dark crumbly soil mixed with small pieces of organic material. It should have a sweet, earthly smell.

Note: Add compost to plants and gardens by mixing it with the regular soil.

QUESTIONS FOR THE CHILDREN:

1. How did the compost look after week one?

2. How long did it take for the temperature in the bin to rise?

3. How does the compost smell?

APPLY:

Try growing beans or other seeds in pots, some filled with sand and others filled with a mixture of sand and compost.

1. Compare how well the seedlings grow.

2. Discuss the plants' need for nutrients and water. Sand is a poor nutrient source and does not store water.

3. When compost is mixed in, both of these needs are better met.

Note: Gardens can be enriched using compost.

Special Words

Biodegradable – able to be broken down by microorganisms into simpler forms.

Collect – to bring together

Compost – a mixture of decaying leaves, vegetables, manure or other organic matter, used to make the soil better for gardening.

Decompose - to rot, or break down.

Enrich – to improve or make better by adding something

Environment – the air, the water, the soil and all the other things that surround a person, animal or plant.

Garbage – food and other things that are thrown away.

Landfill - a place where unwanted materials are dumped, compacted, and covered with dirt.

Pollute - to make a thing or a place dirty, unclean or harmful

Produce – things that are grown or made

Recycle – to collect a waste product and use it in making the same product or another product.

Rummaging – to search completely by moving things around

Trash – everything that is thrown away or considered worthless, with the exception of food.

Waste – Material that has been thrown away or is left over

The Compost Heap

```
G H W R P J Y T R S L L R D D D N R
X R L O Q F S V V X E G E Y U F T U
P J A N R O T P E V V Q I S K O Y X
C O F S P M M Q O B T J N C J H P S
Y D L M S X S H N Z H E X T N D R Q
T E O L H D S A Y L D V W G E I E T
P C B H U N N D W R S K Q A L E C F
E O G Y S T J R A C D I N R L D Y Y
S M C R R H E G S J D A B B X X C D
R P V S G O X S T G P E A A U L L C
Y O F D M U R L E X G T G G S E E S
I S Y K U G S Y Y R E C S E Y A A B
V E H M M O C E U G J D H Z C V A Z
Z W P V U I X W E E M R Q T F E P P
U F Z F I X C V R F F I X J H S A T
A J Y K D T S U P L A N D F I L L L
Y P F P C O L L E C T E D J X B D X
B E E N V I R O N M E N T C D P S X
```

GRASS   SHOVEL   POLLUTES
WASTE   COMPOST   COLLECTED
WORMS   GARBAGE   DECOMPOSE
GARDEN   RECYCLE   VEGETABLE
LEAVES   LANDFILL   ENVIRONMENT

Made in the USA
Middletown, DE
31 March 2023

27851491R00020